Logical Thoughts at 4:00 A.M. A 99.99% Fact-Free Book

OR

Son of *How To Become an Instrument Engineer*
You and Your Aging Boss
99 Ways to Mount Your Instrument *
A Unique Step-by-Step Plan to Pin Point Your Goals and Make Your Dreams Come True as an Instrument Engineer
The Sensuous System *
Rekindling the Desire for Instruments *
Nurse Greg and Nurse Stan's Guide to Good Instrumentation **
The Virgin Engineer (the New Graduate) *
Transmitter Intimacy *
How To Keep Children You Love Off Instruments
What To Do When Your Boss Won't Change
Old Instruments Are Not for Sissies
If I'm so Wonderful, Why Am I Still Poor?
Uncommon Wisdom— Conversations with Remarkable Managers (A Very Short Section)

* Has little to do with what is in the book but is the result of a request from the editor to make the material more sexy.
** Designed to reduce the number of times we are called doctors.

G.S. McWEINER
Illustrated by Ted Williams

Copyright © Instrument Society of America 1991

All rights reserved

Printed in the United States of America

No part of this publication may be reproduced, stored in a retrieval system, or transmitted, in any form or by any means, electronic, mechanical, photocopying, recording or otherwise, without the prior written permission of the publisher.

INSTRUMENT SOCIETY OF AMERICA
67 Alexander Drive
P.O. Box 12277
Research Triangle Park
North Carolina 27709

Library of Congress Cataloging-in-Publication Data

McWeiner, G. S.
  Logical thoughts at 4:00 A.M. : a 99.99% fact-free book / G.S. McWeiner ; illustrated by Ted Williams.
    ISBN 1055617-332-6
  1. Engineers—Social conditions—Humor. 2. Production engineering—Humor. 3. Corporations—Social aspects—United States—Humor. 4. Engineering—Social aspects—United States—Humor. 5. Technology—Social aspects—United States—Humor. I. Title.
PN6231.E58M37   1991
818.5402—dc20   91-35256
                                                             CIP

ABOUT THE AUTHOR

Contrary to public rumors, G. S. McWeiner is not a new meal at the golden arches but is a hybrid author who combines the unique features of two fellows who wish to remain anonymous. He brings to the table over fifty-five years of experience and a lap top computer. The release of the book is timed for the early retirement of one Fellow, and, possibly, new career opportunities for the other Fellow. The book was completed on Paradise Island after many inspirational thoughts such as, "If I was finished, I could check out the topless (non-male) sunbathers." Special thanks go to Carol, Ellen, and Lisa for the endless supply of mangos and rum.

Contents

Preface ix

Instruments Do Not Always Play Music or Introduction: Why Do We Write This Stupid Stuff 1

The History of Instrument Engineers in America—Part II 7

Page Two 15

Yore—As in "Days of..." 19

Neat Things To Do in a Big Corporation 25

Ursa Major, Ursa Minor, and Ursa Mama or A Batch of Stars 29

Self-Test of Batch Controller Configuration Knowledge 35

Elvis is Alive and Well and Living as an Instrument Technician in Indiana 41

Ethics 45

Hot Water 47

Books We Really Need 51

Logic and Order 55

An Engineer's Fantasy	59
Change	61
Spaces	65
Instrument Engineers Adventurers' Club	71
A Pie in Every Cubicle	75
The Top Ten Terrifying Thoughts of Instrument Engineers Just Before They Fall Asleep	79
Electrons	81
Some Wonderful and Nonwonderful Things Computers Will Do	87
How About Giving a Talk?	91
Believe It or Don't	97
How Many Fellows Does It Take to Pick Up a Recalled Car?	101
The Uffda List	103
Phone Tag for Fun and Profit	107
Gerry's Idioms	109
Bad Rap	111
Top Ten Reasons for Buying This Book	117

"Do you swear not to rock the boat of middle management, always try to guess what your manager wants, fear his wrath, and be at his beck and call?"

PREFACE

While there is technical stuff in this book, technically it is not true. There are no "how to" hints, but there are, buried in the book, messages on various levels. Some have extraordinary implications about the health of technology in the United States.

On one level, the material is merely humor derived from absurd experiences in Corporate America. There is value in loosening up and laughing. We can get too wrapped up in our technical problems. We can get too serious and lose a sense of perspective and our humanity, both to the detriment of ourselves and society.

On another level the material points to the insidious and proliferating distractions that prevent the engineer from achieving the high degree of technical accomplishment that is possible today. Liability issues, the unfriendliness of so-called "smart" systems, the quest for the mythical "design without engineering," the emphasis upon appearance rather than substance, and the sacrifice of long-term benefits for short-term objectives are drains on the lifeblood of technology. The engineer is in the enviable position of being able to do something significant to improve the quality of life through the application of emerging technologies, to produce new products, to rehabilitate ecological systems, and to protect the environment. The question is not whether the engineer *can* do it, but *will* the engineer do it? Will he or she focus on socially responsible goals and be given the freedom and resources to accomplish them? Will Corporate America do the right thing or continue on a new path of capitalism that shows a disturbing lack of empathy and foresightedness?

Thus, there is another message level that expresses an even greater concern about the misuse of capitalism. The viewpoint is not one obtained from any education in business principles, but grows from reflections on world events and the observed effect of business goals on engineering and manufacturing in America over the last decade.

The emergence of capitalism as the time-tested victor in the battle of economic systems has rightfully confirmed it as the best for overall prosperity. The basis of capitalism—the desire for money and the associated stature and power—is overwhelming and pops up no matter how much it is suppressed (e.g., the black market). While capitalism is not benevolent by nature, the checks and balances between a free economy, educated consumers, and social consciousness help spread the wealth. Until recently, most wealth in industrial economies was accumulated based on improvements in manufacturing that was accomplished according to the rules of supply and demand. Providing a better product was a way of earning more money. Total quality emerged, and the customer and the manufacturer began to work together in a mutually beneficial relationship. But stock manipulation, take-overs, junk bonds, and the milking of manufacturing, equipment, and institutional assets also appeared as a way of making money. The bankruptcy of once viable corporations and the failure of previously prosperous savings and loans institutions, banks, and insurance companies are flagrant examples of the disturbing new trends. Not as well recognized is the gradual deterioration of the manufacturing base in America due to the misapplication of financial goals, the most widespread of which seems to be the obsession with short-term return on equity. It can be used by stock owners, brokers, and executives for immediate financial gain, since the present stock price greatly depends upon the current return on equity. A short-term increase is obtained by working on the denominator. Operating units are sold, capital expenditures are reduced, and income is used to buy back stock to reduce equity. This inappropriate use of a financial goal tends to tear at the heart of the culture of a company. Rigid orders prevail, flexibility is reduced, the *perception* of what you are doing is more important than *what* you are doing, and risk taking and innovation are discouraged. On the other hand, a long-term increase in both the return on equity and the total cash flow is achieved by an inspirational strategic goal—work on the numerator—and by providing resources for the implementation of new technologies. The article "Strategic Intent" in the May-June 1989 issue of the *Harvard Business Review* eloquently details the need and the value of focusing on the strategic intent rather than on financial goals.

The diminished stature and the deterioration of manufacturing is not noticeable from the outside. The appearance of well-being is maintained through glowing financial reports, optimistic views of the discovery of miraculous new products, beautiful facades, artistic landscaping, and impressive meeting rooms. Meanwhile, plant equipment, technological expertise, the ability to sustain existing products, and the ability to commercialize new products is deteriorating. The stock market, perhaps unknowingly, has become the perpetrator of a grand illusion.

We can blame the brokers, investors, and executives who are ingeniously exploiting the system, but more likely we should blame ourselves for setting the climate for economic disintegration and not demanding more socially responsible longer-term goals. It is an expression of our own insatiable appetites, our "use up and discard" culture, our movie set mentality, and our need for instant personal gratification. It is time to wake up. We are nearing the end of an era of manufacturing in America and proceeding toward an economy based on invention, service, and financial manipulation.

Humor has a way of breaking down barriers and making people aware and (hopefully) ready to communicate about particularly difficult and controversial issues. The first requirement in changing ourselves is the ability to laugh at ourselves. So relax, enjoy a few chuckles, exercise the right side of your brain, let your soul emerge, and then set up an appointment with your CEO to work on long-term improvements in the numerator.

Oh, and tell your CEO that G. S. McWeiner sent you.

HI! I'M AN INSTRUMENT DESIGN ENGINEER!
REALLY? DO YOU PLAY THEM, TOO?
THE FLUTE....
LITTLE THINGS YOU MASH ON
END YOU BLOW INTO
LONG SHINY TUBE
END SOUND COMES OUT OF
?
ISA

INSTRUMENTS DO NOT ALWAYS PLAY MUSIC
OR
INTRODUCTION: WHY DO WE WRITE THIS STUPID STUFF?

Our first book, *How to Become an Instrument Engineer*, was written with the belief that it would be read and understood only by instrument engineers. Soon after publication, it became obvious that all kinds of people were not only reading our book, but understanding and enjoying it. We received compliments from college professors, MBA's, sales people, psychologists, and even people with engineering degrees (notice we did not say "engineers"—all engineers do not have degrees and all people with engineering degrees are not engineers—some are lawyers and others are managers). In all honesty, we must admit that no one in a management position ever complimented us directly or in public. Some are even under the impression that we are giving out company secrets and/or criticizing the way large corporations are operated. One of us was even accused of having a COMPLETE CONTEMPT FOR ALL LEVELS OF

Logical Thoughts at 4:00 a.m.

MANAGEMENT! Fortunately, we were prudent enough not to ask him how he knew.

So we decided to write another book with the following criteria:

- There would be little or no technical content.
- As many people and organizations as possible would be "evaluated." The book would have so little redeeming value that it would not have to be reviewed by our management, other technical experts, or the publisher.
- By the use of balderdash, various vintage, veiled, virtual verities and cartoons, we would begin the education of the world.
- Whoever bought his own copy of this book and heeded its message would surely never despair at not reaching the lofty heights of middle management.
- We would demonstrate to the world how well some engineers can communicate.
- A full chapter would be devoted to ethics and dress codes. One example asks the question, "If your future mother-in-law throws you out of her house one week before the wedding, do you have to have her son as an usher and allow him to wear a tuxedo?"

WHAT ARE INSTRUMENTS?

There are two kinds of INSTRUMENTS.

Type "A" is the kind that plays tunes. Examples include the fagott, the sackbut, and the kazoo. Stereo receivers, compact disc players, and equalizers do not fall into this category.

Introduction

Type "B" is everything else except stereo receivers, compact disc players, and equalizers.

This book is more interested in Type "B" even though the authors are more interested in Type "A."

If you understand the last five sentences, you can sell this book to your boss. He needs it more than you do.

HIM NEED HEAP GOOD HAIRCUT...
BIG CHEESE
...NO INDIANS, BUT THERE ARE A WHOLE LOT OF MOVING BUSHES... WOULD YOU LIKE TO PLAY A GAME OF MUTANT NINJA CAVALRYMEN TO KILL TIME?

The History of Instrument Engineers in America—Part II

Since no one that we know has ever written a history book on Instrument Engineering, we are going to make one up. If anyone can prove that anything in this book ever really happened, or really never happened, we will publicly apologize the next time one of us is asked to give the keynote speech at some annual convention. With that out of the way, we can begin.

We start with Part II because not much happened in Part I or at least not much was recorded in the annals of Colonial America. There are some vague references to the use of gravimetric feeders for the corn crop of the early Pilgrims, but the existence of a large central staff of instrument engineers at Williamsburg is considered to be just a rumor started by the earliest known drafting department. Instrument engineers came out (or more directly, first found their way out) of the closet in the late 1700's. It is a little known fact that they were the colonial army's secret weapon during the American Revolution. Most of their activities were shrouded in secrecy to keep the British from formulating a counter measure. What the Americans didn't know was that the British were developing their own contingent of instrument engineers but were still trying to define the crumpet and spot-of-tea details. Who knows what might have happened if the war had lasted long enough for the British to learn how to calibrate a d/p cell. We might have had Margaret Thatcher instead of Nancy Reagan running our country and the organization "Astrologers for a Stronger America" might have been relegated to just "horsing" around at a ranch in California.

Thanks to the effort of Maggie Meter (Mag, for short), the solitary Reagan-era instrument engineer, some of the contributions of instrument engineers as revolutionaries and pioneers have come to light. The details were revealed to Mag after two days of non-stop start-up coverage when the ghost of her great grandfather, Head Meter, personally visited Mag and set the record straight.

PAUL REVERE, AN ALARMIST IN THE TRUE SENSE

Unbeknown to most modern technologists, the problem of excessive nuisance alarms was solved by Paul Revere. Before Paul, the role of warning the revolutionaries of British military movements was relegated to Bob Disregard. Bob suffered from poor eyesight and was prone to exaggeration, especially after a long evening at the tavern. If Bob saw a red coat while lying on the floor, he would struggle to his feet and stagger through the streets yelling, "The Red Coats are coming!" Since red velvet coats were fashionable, local inhabitants learned to shut their windows and wear night caps to muffle the sounds of Bob's nightly alarms. A particular problem around Christmas time, Santa quickly learned to steer clear of Bob. Paul solved the problem by telling Bob that the British would actually be wearing dayglow orange suits. Paul then assumed the foe-warning role and eliminated false alarms by adding intelligence to the redcoat detection logic. Paul correctly reasoned that a red coat by itself represented insufficient data and decided to sound the alarm only if these redcoats were marching, carrying guns, playing drums, and doing all the typical soldier things of the day. Paul was so revered by the city of Boston that a line of pots and pans were named after him, much to the dismay of Ken Tupper.

THE DEPTH OF THE INSTRUMENT ENGINEER CONTRIBUTION

It was fortunate that George Washington had an instrument engineer aboard when he crossed the Delaware; otherwise, he might never have made it to the other side. As you know, the crossing was made in fog as thick as Martha's pea soup. What you might not have realized is that George relied upon the use of level sensors procured by his ace instrument engineer, Flippy Float, to determine whether he was approaching the shore or just going south for the winter. The level sensors were sticks with graduations marked in metric units (they were supplied by the French). George had no idea what the numbers meant but wisely recognized when they increased or decreased (the same mode of operation largely in use today).

The application was not without its start-up problems. The first two sticks failed immediately, despite the fact they were different in design and manufacture. The graduations on the first brand of stick selected were made with a colorful but water-soluble dye. The second type had an extremely lightweight and streamlined construction but broke when it hit a rock. When queried by George as to whether these sticks had been tested, Flippy replied that they had been extensively inspected for compliance with the specifications by the warehouse clerk upon arrival and had their calibration thoroughly verified by the instrument technicians. When George rather heatedly asked, "But did you poke them in water?" Flippy answered, "No, we couldn't find a bucket big enough in the shop, and it was too cold to try them in the river." Flippy was about to be flipped in the river to find out how cold it was when he produced a third stick. Flippy said, "We can try this one, but its specifications are not as good and it is rough, rather thick, and quite drab." George then said one of the many things that made him great: "We are not here to win any beauty contests; we will leave that for the British." The third stick was rugged enough to withstand the operating conditions, including a blow to the seat of Flippy when he told George he bought a hundred of the first two and only one of the last type because it was more expensive.

The application was ruled a landmark success. In a brilliant stroke of military genius, George mailed the remaining defective sticks to the British military command marked "TOP SECRET" with a London return address. The maneuver is credited with the running aground of the British fleet in a subsequent battle as evidenced by the numerous unmarked and broken sticks that floated ashore.

THE ANSWER IS BLOWING IN THE WIND

The North's use of hot air balloons in the Civil War for Confederate troop observation got off to a rocky start. West winds blew launched balloons so often to the coast that Federal scouts took to wearing bathing suits. A rebel spy designed a weather vane for the North that used strategically placed magnetic material that caused the instrument to always indicate a south wind. Consequently, the Federal scouts decided they needed to get south of the Confederate troops to launch their

History of Instrument Engineers in America—Part II

balloons and disguised themselves as clowns headed to a circus in Atlanta. However, when asked by suspicious rebels the first name and middle initials of General Lee, they incorrectly guessed Jim B. for Jim Bob. After a careful review of the clowns, the Confederate officers decided their cause could be best served by letting the scouts continue on. As northern winds settled in for the winter, the scouts were last seen headed for Florida.

AN ANALYTICAL LOOK AT CUSTER'S LAST STAND

The first known application of an analyzer was by General Custer. A shipment of Indian detectors, complete with simulated Indian figures, arrived in time for Custer's instrument engineer to try them out before Custer's intended tour of the Little Big Horn area. Custer initially balked at the application because the units required a protected environment, and a portable analyzer house would be expensive and would slow him down. However, Custer reasoned that the use of Indian detectors would allow him to cut the troop requirement and associated costs and get an outstanding rating for his next performance review. But Custer's ruthless nature would come back to haunt him. The lack of Indian prisoners meant the closest thing to a real Indian was the figures supplied with the detectors. The instrument engineer dressed up his dog as an Indian, but the unfortunate creature was immediately shot by Custer and added to the body count reported to Washington.

The instrument engineer calculated the best he could do was place the Indian figures about a thousand feet away and hope that they would be equivalent to real Indians (about twenty times larger) that were almost four miles away. The test was a success, and despite warnings

11

Logical Thoughts at 4:00 a.m.

by the instrument engineer that simulation is never as good as the real thing, Custer, anxious to add to the luster of his quarterly report, proceeded with the first application. The result of Custer's last stand has been retold in many ways, most of which depict Custer as heroically kneeling and firing his pistol surrounded by dead soldiers and live Indians. Eye witness accounts by Indians state that Custer was actually in the analyzer house at the end of the battle, furiously trying to get an on-scale reading.

NEW COMPUTER CONTROL CENTER
TECH·NO·TANK
HOW DO THE OPERATORS LIKE IT?
...THEY THOUGHT WORKING WITH THE GRAPHICS WAS NICER THAN DEALING WITH REALITY UNTIL THE DISPLAYS SAID THAT EVERYTHING WAS FAILING...

Page Two

Don't begin with writing on the first page. At some later date you may find something very important to say first.

BROKEN

    1a1. If something doesn't work, is it broken?
    1b. If something is broken, is there only one thing wrong?
    1c. If there is more than one thing wrong, is it brokens?
    1d. If brokens isn't a word, is it broken, broken, ...broken?

Back to 1a:
1a2. If something is dumb or really bad, is it broken?

Example of 1a2:
A company builds an automobile with an engine that is too small, an incompatible transmission, windows that leak, an inadequate electrical system, etc., etc. Should the car be called

    a. A lemon?
    b. A pile of junk?
    c. Broken?
    d. A junk pile of broken lemons?
    e. A broken lemon?

Logical Thoughts at 4:00 a.m.

Questions about 1b:

When someone goes to fix something that is not working, he/she looks for something that is broken. As soon as they find a problem, there is a feeling of accomplishment. The next step is to repair or replace the defective item. There is now an assumption that the thing is no longer broken—it is fixed. However, if one part breaks or wears out, it may cause other parts to break or wear out, which may cause other parts to break or wear out...etc. So many parts may be bad that it isn't worth fixing the whole thing.

1a3. If a person does something that causes a mechanical device and/or computer to malfunction, is it broken? If someone shuts off the power to a computer or doesn't put gasoline in a car, are they broken?

What causes the most things to malfunction?
  a. Broken or worn out parts
  b. People who do stupid things
  c. People who don't do the right things
  d. Stupid people who use broken or worn out parts incorrectly
  e. People who use the wrong thing because they didn't understand the job requirements or have the correct part
  f. Parts is parts

I GUESS HE NEEDS THE SPACE TO DEVELOP A VISION FOR THE LONG RANGE PLAN...
INSTRUMENT CATALOG
BOOK OF BIG WORDS
AMERICA
INSTRUMENT CORNUCOPIA
TECHNICAL STUFF
ISA
W. VIRGINIA POTTERY
PAPERS
ISA CATALOG
LOGICAL THOUGHTS AT 4^{00} AM
WALL STREET JOURNAL

Yore
As in "Days of..."

Recent Yore

A number of years ago in a little town on the East Coast, there lived a very unhappy man. He had convinced a group of friends and neighbors to invest their money in a joint venture that would make them all rich and famous. He had told them that only twenty months after they gave him $89,327.62 plus or minus 15%, he would produce two batches per day (42 gallons per batch) of a compound that would provide Leather Improvement by Artificial Rolling. Yes, you guessed it. The code name was "LIAR." However, since he was basically an honest man, he wanted to do the correct thing.

Once this man (whom we will lovingly refer to as our "Project Manager" or "PM") received the funds to proceed, he realized that he was in deep egesta. He had spent all his time and effort in convincing the investors that he was truly talented. The PM had not consulted with the proper people to fully develop the manufacturing process. He knew that a building was required to store the raw materials, produce the product, store the product, provide offices for the executives, and have a conference room for the investors and a broom closet for the janitor. But he had not calculated the required size of each space.

No lunchroom was needed since the workers could eat at the saloon across the street. While they were at lunch, they could also take care of their biological needs. This PM was a master at making the most out of limited

Logical Thoughts at 4:00 a.m.

resources, which included his intelligence. What to do? Within a relatively short time, he developed his LRP (Long Range Plan).
1. Hire new college graduates with grade point averages below 2.5. They would work hard and long for very little money. They would be grateful that he had hired them because the large corporations would not even talk to them.
2. Get as much free assistance as possible. Examples included allowing the local tank supplier to size, support, and build equipment. The ice supplier could build the ramp to get his cart into the building.

The PM knew that he would have to write periodic reports with tables and graphs. The problem was that he didn't want to take the blame for possible delays and additional costs. He needed a scapegoat. He needed to have someone on the payroll who did something that nobody else understood. He needed a person who could not communicate with other people and so could not defend himself. The answer did not come easily.

To make a long story a little shorter, he went to the local university where he walked up and down the halls looking for the strangest, most noncommunicative students. They turned out to be physics majors. Unfortunately for the PM, there were very few physics majors and they all wanted Ph.D.s, refused to work, and probably couldn't do anything that even sounded remotely useful. The second choice was an electrical engineer (or EE). (Those were the days when the students and professors in any university spoke words in the local language, that is, not saying more than four words in a row that made any sense.) Anyway, an EE interviewed and was hired. He was told to buy a suit that fit and report to work with all his used text books and slide rules.

QUESTION: Does anyone ever use their old textbooks?
ANSWER: Wait. Read on.

Yore—As in "Days of..."

On the day our young engineer reported to work he was given the following information and instructions:

- A schedule of when he was supposed to complete the different phases of his work.
- A capital budget equal to 2% of the total cost of the project.
- A manpower estimate equal to 25% of his capital budget.
- Instructions to provide "devices" that would correct the shortcomings of any piece of equipment supplied by anyone else associated with the project. In addition, once the plant was started up, these devices were to prevent the operators or supervisors from doing anything that would damage the plant or hurt someone.

Believe it or not, our young EE did everything he was asked to do— and more. In other words, he saved the PM's hiney. The project was a success. The investors all got rich. The PM received a healthy bonus and our hero received a decorated ash tray for his desk and a 4.5% increase in salary. But, most important, he received a new title—"Instrument Specialist." After 35 years of dedication and long hours, he retired and went to work for a local consulting firm with the title of "Control Systems Specialist." This allowed him to take long weekends to go fishing in his pickup truck.

"The prototype professional for the 1990s"

Neat Things To Do in a Big Corporation

In the presentation of this material to ISA Section meeting attendees and short course students, people have laughed and related to the same situations regardless of their company of choice. It seems that some themes in Corporate America are just naturally a rich source of humor. The following things are arranged in increasing order of neatness with the hope that the first step to coping with a predicament is to laugh at it.

10. Watch the number of management courses go up as the level of technical courses goes down. Spend all your working hours attending management courses. Go to a stress management course when you can't solve your technical problems.
9. Watch the number of site administration employees go up as the number of engineers goes down. Make sure there are international symbols for staircases, coat rooms, and meeting rooms so that foreign visitors don't relieve themselves in the staircases or try to have meetings in a coat room.
8. Watch the number of vice presidents go up as the number of engineers goes down. Add them to your Christmas list. Drop in for coffee and chit-chat.
7. Watch the number of slabs of granite and colored glass go up as the number of engineers goes

Logical Thoughts at 4:00 a.m.

down. Donate one in memory of an ex-engineer. Look for one with your name on it.

6. Create a new department and post the announcement on the bulletin board. Restructure it yearly. Promote your friends. See how long it takes for someone to realize it has no purpose.
5. Collect swell gifts from stockholder meetings. Give them as presents to your friends. Look for new friends.
4. Play Russian Roulette with the food machine that replaced your cafeteria. Use blindfolds, nose pins, and hot sauce to prevent sensory-induced illness. Give a prize for the highest cholesterol test result.
3. Develop maze games to liven up an otherwise dreary generic cubicle existence. Place summer employees in the center and see how many days it takes them to find the exit. Make them wear numbers and place bets. Place doughnut piles and coffee pots in strategic places to stave off starvation and dehydration.
2. Hug your boss after your next performance review. Plead for an encore and an autographed paycheck. Name your children after him or her.
1. Obtain your CEO's favorite book. Read passages backwards to look for hidden meanings. Add illustrations.

...QUESTIONS?
MANAGING A DIVERSE WORKFORCE
1990...
1 BEAR
2010...
256.3 BEARS

Ursa Major, Ursa Minor, and Ursa Mama
or
A Batch of Stars

Our civilization is moving in some direction. Whether that direction is right or wrong probably can't be answered by an engineer or any kind of physicist. However, since it is moving and/or changing, things need to be brought up to date.

The first thing is the THREE BEARS.

No Goldilocks—just the THREE BEARS.

Mama Bear was getting a lot of complaints about the porridge bowls being empty and the delays in getting the porridge on time with consistent quality. The temperature and mineral content of the well water was always changing. The solid porridge was hard to measure on cold, dark mornings. The cooking time varied because of the wood stove and, worst of all, Mama Bear had to pay attention to what was going on.

In order to save her marriage and keep Baby Bear from running away and becoming a rock musician, she began an investigation of hot cereal preparation in America. After weeks of talking to local vendors, reading technical magazines, and attending courses at the local university, she developed a plan to solve the problems one at a time.

Logical Thoughts at 4:00 a.m.

The plan included plotting the effects of the solutions on her family's behavior.

By installing a large glass-lined storage tank, ion exchange bed, and circulating pH control system, she was able to stabilize the temperature, acidity, and mineral content of the well water. Instead of trying to measure small quantities of solid ingredients on cold, dark mornings she increased the size of the batches and made them on sunny afternoons. She bought and installed a special interface card in Papa Bear's home computer. By measuring the temperatures of the pot and the fire, the PC could calculate the correct cooking time. This allowed Mama Bear to watch the soap operas while the pot cooked. When the computer signalled that the batch was complete, she packed it out in individual plastic containers and quick-froze them to capture the "fresh-cooked flavor."

At bedtime, she took two portions from the freezer, put them in the microwave oven, and programmed the timers. Just in case there was a variation in the schedule, the timers had an input from the water flowmeters on the toilets, which were a check on how things were going. When Papa and Baby came down to breakfast, the porridge was ready, hot, and consistent. If this was an old-time story, it would end: "And everybody lived happily ever after." But this is an updated version; there is a little more to tell.

Papa was happy for a few weeks but soon grew frustrated about not being able to complain at breakfast time. Baby stayed at a friend's one night and had three pizza specials delivered for dinner. Being little bears they couldn't finish all the pizza that they had ordered so they left the remains on the kitchen counter. When the little bears got up the next morning, the friend's mother was still asleep so they ate the leftover cold pizza from the night before. Baby Bear decided that it was "cool" to eat cold leftovers for breakfast. So now in the morning Mama

Ursa Major, Ursa Minor, and Ursa Mama

Bear had to deal with an angry Papa Bear who had nothing to complain about and a Baby Bear who made her sick eating mixtures of incompatible foods at the wrong temperatures.

Mama gave up her automated food preparation and went back to college to become a Mental Health Professional. Papa gave up breakfast to lose weight. Baby became a normal, obnoxious teenager. Mama received her Master's Degree in Social Work and went to work for $11,000 per year dispensing welfare checks and listening to more complaints than she had heard while serving inconsistent porridge. Being a little smarter than the average bear, she resigned her position and founded the first chain of automated, fast, leftover, cold pizza and Chinese food restaurants. She located all her restaurants across the street from high schools in California. Once she raised some additional capital, she planned to expand to the East Coast to take advantage of another large group of weirdos. Papa became the corporation's chief legal advisor in charge of their Diverse Work Force, and baby became the "Lack of Quality Expert" (remember, they were feeding high school students).

...WHEN WILL THE CONFIGURATION BE DONE ?
Distributed Control System
REP.
...AS SOON AS MY ENGINEERS GET BACK FROM VACATION...
APEX CHEM.
HaHAHA! HEH! HEE! HO! HA! HE HE HA! HEEE...
LOADING DOCK
GOTHAM CITY'S REHABILITATION UNIT FOR CONFIGURATION ENGINEERS

Self-Test of Batch Controller Configuration Knowledge

1. How do you estimate the configuration costs of a batch controller in your distributed control system (DCS)?
    a. Carefully figure the number of hours required to complete the controller definition, recipes, operations, interlocks, and data base.
    b. Ask a configurer how much time was spent on a similar job (if he's still coherent) and double the number to allow for meetings, management courses, and DCS enhancements.
    c. Choose the largest number you can get by your project manager.
2. How do you learn to configure a DCS batch controller?
    a. Rent a U-Haul trailer, fill it with DCS instruction manuals, take a leave of absence, and read every page.
    b. Go to a DCS configuration school and hope your first job consists of emptying and filling a couple of tanks with water.
    c. Find the simplest and safest application with the lowest profitability, visibility, and priority. Convince your boss you are the only one who can configure a batch controller for

it and beg the most knowledgeable configurer around to act as your tutor.
3. You have a hundred virtual points. This means you are
   a. a virtuous person.
   b. a graduate student studying data highways.
   c. headed for a communication breakdown between your batch controller and the highway and between you and your boss.
4. You are convinced that you don't need real-time simulation to test your batch controller. This means you are
   a. good at hiding start-up costs.
   b. starting up a plant in Hawaii.
   c. scheduled for early retirement and see a bright future for fixing configuration problems as a contract engineer.
5. An operator's console has no active alarms. This means you have
   a. figured out how to use every feature of alarm group states to suppress unnecessary alarms.
   b. perfect instruments and operators and an especially smooth-running process.
   c. forgotten to download the console.
6. Your configuration for the batch controller fits in one small binder. This means you have
   a. demonstrated outstanding conciseness of code.
   b. an application of emptying and filling two water tanks.
   c. forgotten to do the data base.
7. You have 1001 syntax errors. This means you have
   a. an incredibly complex application.
   b. a computer virus.
   c. forgotten a comma.

Self-Test of Batch Controller Configuration Knowledge

8. You think you have the best of all possible batch controllers in the best of all possible distributed control systems. This means you
   a. have installed systems of all the major competitors.
   b. have suffered a time warp back to the era of Voltaire.
   c. are a sales engineer.
9. You write fan letters to the technical writers of the documentation for your batch controller because you are
   a. an avid reader of technical writing.
   b. ardently collecting binders as a hobby.
   c. a former employee of the DCS rep making big bucks as a contract engineer doing configurations.
10. If you met the originators of the first configuration workstation and error message set, you would most like to
    a. thank them profusely for promoting creativity.
    b. ask their planet of origin.
    c. tie obsolete workstations to their waists and take them for a swim.

HEY... WHY DOES THE "KING" HAVE LOAD CELLS ON HIS BED?
HE HAS SECRET ASPIRATIONS OF BECOMING A "BIG MAN" IN INSTRUMENTATION...
ASK ME about INSTRUMENTS
LOGICAL THOUGHTS at 4AM

Elvis is Alive and Well and Living as an Instrument Technician in Indiana

Elvis has been spotted (usually on the third shift) gorging himself on stew, red beans and rice, ribs, slaw, and jelly doughnuts or, in other words, the typical fare of any control room's kitchen. He is a great hit with the operators, playing the guitar and spinning tales about various feats performed by corporate instrument engineers. The more he sings, the more food he gets. Needless to say, he has become a BIG man around the plant. Some of the stories become the lyrics of potentially great hits, like "Instrument Shop Rock."

Elvis was exposed to instruments as a baby when his mother would read bedtime stories from *INSTRUMENTATION TECHNOLOGY* (now known as *INTECH*). He especially liked the articles in which manufacturers described in glowing words the wonderful features of their products. Elvis would get lost for hours in this fantasy world where everything was perfect. He collected glossy color photos of control valves (green and red) and colored for hours the outlines of valve bodies and trim in sizing catalogs. Reliable sources say that Elvis's first words were "split range," and that while other toddlers were stacking blocks, he was assembling positioners. Instead of going to the playground, he stole away to the warehouses of a vendor of green control

Logical Thoughts at 4:00 a.m.

valves where he fondled seldom seen parts. His interest in instruments expanded to field transmitters, controllers, and eventually pneumatic computing modules whose many bellows and springs were particularly intriguing to the young mind. By age ten, he had the largest collection of instrument catalogs in Memphis, and the walls of his room became a panorama of the diversity of the windblown era of instrumentation. When he found out that some of the most innovative techniques for pneumatic instrument maintenance were practiced in Indiana, Elvis vowed he would someday live there.

Elvis's venture into a career of rock and roll was an accident. He got into music when his mom gave him a guitar and told him it was an instrument. In the process of calibrating it, some ants crawled into his pants and caused him to wiggle his hips and jerk his pelvis. Neighborhood girls caught the act and started to scream. The rest is history.

It was obvious to the close associates of Elvis that he held secret yearnings to become an instrument technician and a Hoosier. It was most evident when he stared spellbound at the meters and dials of recording studio instruments. Sometimes he would lunge at the electronics with a screwdriver and would have to be restrained by his bodyguards. Fans found they could gain their way into Elvis's inner sanctum by brandishing copies of *INTECH* and calendars from various instrument representatives. An actual vortex meter would insure anyone a spot on the next tour until Elvis learned how to calibrate it. He became obsessive in his attraction to instruments and started to poke friends with an RTD and sleep with a d/p cell under his pillow. He had his bed installed on load cells and trend recorded his weight. He was also known to hold turbine windows out the window of his limo to test the strength of various bearing designs. Elvis once went to a K-Mart to buy clothes to disguise himself as an

Elvis is Alive and Well...in Indiana

instrument engineer in order to gain entry to an ISA exhibit, but he was rejected at the door because he forgot to wear a pen and pencil pocket protector.

The lure of an instrument technician job offering in the great Hoosier state of Indiana was just too much for Elvis to resist. He worked out an elaborate plan that involved the construction of a three-hundred pound robot that could gyrate its pelvis and sweat. Unbeknownst to his fans, Elvis's last concert was actually performed by the robot. The robot's batteries ran out the day of Elvis's supposed death. Meanwhile, Elvis started an agent-free existence as an ace instrument mechanic for a rubber chemicals plant near the Indianapolis Speedway.

If your friends don't believe this fascinating disclosure, tell them it is in a book published by the Instrument Society of America and registered in the Library of Congress. If you would like a picture of a d/p cell signed by the "King" himself, send a hundred dollars (cash only) to the author. Allow at least two years for delivery.

Ethics

There are many different kinds of ethical decisions, procedures, behaviors, and manuals.

In addition, people from various backgrounds and levels of formal and nonformal educations have expectations of the people they come in contact with. However, since ethical questions may be part of language, many things are not clearly understood or even realized. In order to stimulate thought, we offer the following:

- If someone is on a business trip, the company is paying all the expenses including meals, and the person uses coupons to eat in a fast food restaurant, does the discounted or full cost of the meal get charged?
- If you get a free ride to the airport, is the company obligated to pay the equivalent taxifare?
- Is it ethical or sexist to criticize a man's dress code when women in the same position can wear anything they want to work? One of the worst examples of this is a symphony orchestra where men all have on the same type of formal wear while the women players have a variety of gowns, pant suits, mannish-looking black suits, etc.
- Is it ethical for a project manager to report the technical performance of the project team to remote managers and directors who do not understand the technical details so that they can make irrational recommendations?

Hot Water

One day a young engineer was called into the supervisor's office. He was told that other engineers were designing a tank with a sight glass in the side so that they could look inside to see what was going on. The problem was that the sight glass would get coated with the material that was in the vessel and become useless. However, if hot water was sprayed against the glass, the coating would wash off. The problem was that there was no hot water in the vicinity of the tank.

The answer was obvious to our young engineer so he said, "Go to the local home supply center and buy a 30- or 40-gallon hot water heater for about $150, like the you have in your house." Our hero lost at least two points for his "smart-schmuck" answer and was told to get serious. Two weeks later, he came back with his $15,000 solution. There were high-speed temperature transmitters, special batch controllers, control valves with positioners, pressure regulators, strainers, pipe, wiring drawings, etc., etc.

The new solution was based on injecting high-pressure steam into cold water in order to heat it up quickly whenever it was required. The noise and vibrations caused by putting steam into cold water are something that should be experienced by every thinking, feeling, human being. It could be described as "the simulation of the beginning of the end of the world as we know it."

At this point we could end our little story with some kind of simple moral like, "Engineers do not know how to make hot water." However, this would not be fair to

Logical Thoughts at 4:00 a.m.

our hero, who had a good idea that was rejected by management. We could blame management by coming up with an even simpler moral like, "Management doesn't like good ideas if they seem too obvious. They are paying those big salaries and they want complete results." No, there needs to be something very profound in this uncomplicated story about water, heat, a young inexperienced person, and a low-level supervisor (which is kind of like a junior officer in the Navy).

(Some time later:)

Is it possible that we can't see the simple moral any better than the supervisor who couldn't understand why a hot water heater from the hardware store might be the best possible solution to the problem?

...WHAT WE REALLY NEED IS FOR A HARVARD MBA WHO IS AN ENGINEER AT HEART TO WRITE THAT ENGINEERS DETERMINE WHETHER A PRODUCT PERFORMS LIKE A MERCEDES OR A YUGO AND ADVOCATE A NEW PERFORMANCE INDEX FOR STOCKS THAT MEASURES THE COMPANY'S TECHNICAL CAPABILITY...

Books We Really Need

Instead of another book on control theory, what we need are books that will really help our careers. We think the promotion to Super Engineer may be just one book away from us. If only we could find the right one. Here are some of the titles we are looking for:

How to Impress MBAs

Creative Time Sheets

Can Computer-Integrated Manufacturing Be Your Meal Ticket?

How To Expense Instead of Capitalize

Appearances Are Everything

Short Cuts to Early Retirement

Fool-Proof Goal Document Verbiage

Emerging Careers as a Meeting Specialist

How To Win Phone Tag

The Best Buzz Words

How To Turn a Cubicle into an Office

How To Look, Act, and Sound Like You Have an MBA from Harvard

The Complete Guide to Special Function Keys

Impressive Graphics for Average Results

Behavior Modification of Your Boss

101 Ways To Compliment Your Boss

Logical Thoughts at 4:00 a.m.

Antidotes for Reading "How to Become an Instrument Engineer"

The Comprehension of Illogical Management Directives

How To Be Visible for Successes and Invisible for Failures

I Don't Want a Dollar Tomorrow, I Want a Dollar Today

Taking the Capital Out of Capitalism

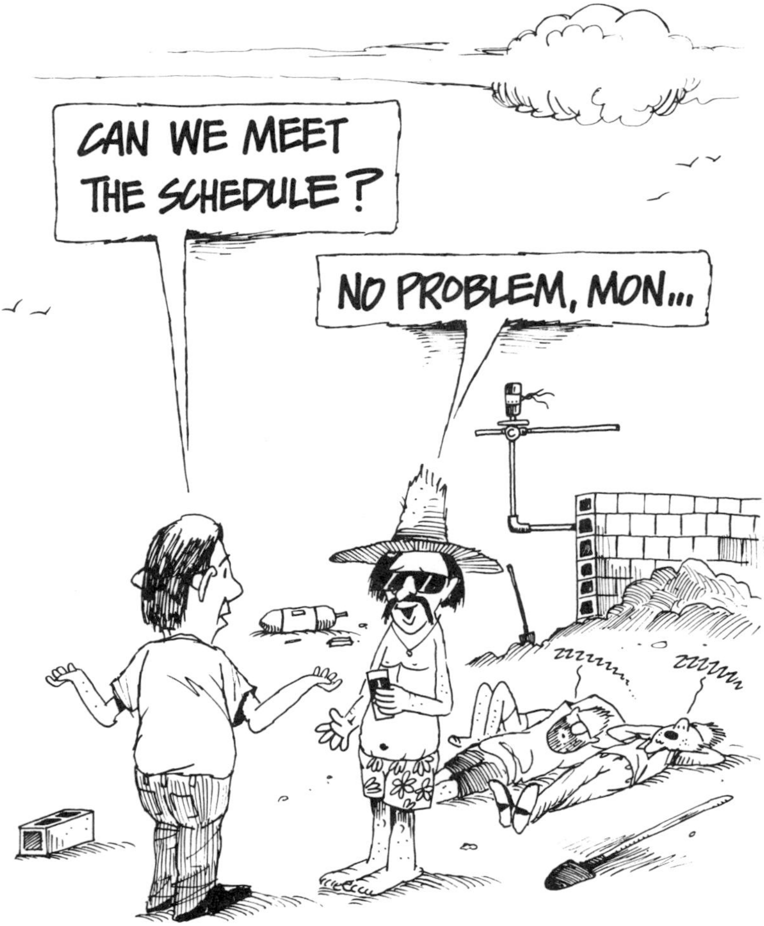
CAN WE MEET THE SCHEDULE?
NO PROBLEM, MON...
zzzzzz...
zzzzzz

Logic and Order

People do jobs or things in a certain order. We say that this order comes from learned techniques and/or experiences. The order may not make any sense or be logical to other people. Even if it can't be explained, or it is explained, or whatever else, they may not want to change the order. If the people come from a different part of the country, a different country, talk differently, or look different, there is a story about why their order is "screwed up."

Here is a story about an American who went into a foreign country and tried to change the order of things.

Once upon a time, a competent engineer went to a foreign company to assist in the design of a new plant. The work was being done by a local engineering and construction company that had their offices in the downtown area of a very large city. Our hero went to work at 8 a.m. on the first morning, expecting to find everyone hard at work. His first problem was that the only person at the office when he arrived was the night guard who did not speak much of any language. About 8:45 a.m., a secretary, who knew that he was coming, convinced the guard to let him in. She showed him his assigned office, shut the door, and left. Since there was nothing else to do, our hero worked on his expense account. At 9:27 a.m., hearing some noise outside his office, he looked out to see 6 or 7 people taking off their jackets and preparing their desks for some activities. A short time later, a pleasant-looking person who spoke perfect English introduced himself as the project manager and apologized for being late because the day before was a holiday and

Logical Thoughts at 4:00 a.m.

everyone went to the beach. He explained that there would be a meeting that afternoon for our hero to meet the staff.

At 4 p.m., the meetings and introductions took place. Plans were made for the execution of the work. At 8:15 the following morning, the American returned to the office eager to begin the project so that he could leave work at a reasonable hour to enjoy the hotel swimming pool. The sleepy guard recognized him and allowed him to enter the offices. At about 8:45, the first secretary arrived. By 9:30 most of the staff had arrived, so the hero requested a meeting. He was told that there was so much to do, nothing could be done with him before 3 in the afternoon. Could it be that almost nothing was done by the people in that company before the late afternoon?

A few days later, the local representative of an international corporation called to request an appointment with the visitor from America. The time suggested was 4 p.m. Our hero suggested 9 a.m. The meeting was scheduled for 3 p.m. and took place at 4 p.m.

During the second week of the visit, the visitor came to work at 8 a.m. He worked by himself until 11 a.m. when he returned to the hotel pool to gaze at the local scenery, eat lunch, and read. By 1 p.m., he returned to work by himself until 3 p.m., when the meetings began.

The moral seems to be that changing the way a country does business is more difficult than changing the time an engineer goes to the swimming pool.

THE REAL REASON OUR STOCK HAS GONE UP IS THE EXCELLENT JOB DONE BY ENGINEERS IN BUILDING, MAINTAINING AND OPERATING OUR PLANTS!
OOOOOO....
AAAAAAHHH....
Dave

An Engineer's Fantasy

What would it be like to work for a company that provided the following?

- An office with a door
- A classical and rock music sound system
- A secretary that files papers and books
- A maintenance department that cleans the place
- The authority to make technical decisions such as the purchase of a waterproof computer system
- On special occasions like anniversaries, serve Chinese food and pizza instead of "sheet"cake and coffee
- The freedom to select your own travel agency and car rental company
- No personnel or "human resources" department
- A free-form dress code
- A company-operated cafeteria and store with higher quality and lower prices than McDonald's and K-Mart
- A cafeteria menu that would include:
  a. pizza steaks on hard Italian rolls,
  b. submarine sandwiches or hoagies,
  c. Dr. Brown's black cherry and cream soda,
  d. Tasty Kakes, and
  e. hard shell crabs and beer on Friday
- An expense account that permits carrying over cash and taking people out to dinner

Logical Thoughts at 4:00 a.m.

No MBA's, accountants, or lawyers would work in any management position. To qualify for a management position, all applicants would have to write a 30-page paper on "Management by Authoritative Dictatorship" (commonly known as MAD). This paper would bring out the potential manager's true feelings. It would become a permanent part of his/her personnel file for all to see.

Change

If a person or engineer learns to do a job in a certain way and it works, there is a tendency to keep doing it that way without even considering alternative ways.

The reasons why the method or procedure works is of no consequence. Even if the method or procedure is more complex than is necessary and it works, it still becomes the model for future projects.

There is no incentive to change.

The reasons that are given for not changing:

- It might not work if changed.
- Why risk change?
- What do I get out of it?
- If you didn't change it and it doesn't work, it's not your fault.
- If you change it and it doesn't work, it's your fault.
- We like it the way it is.
- They like it the way it is.

Suppose you got points for CHANGE.

Call it TRYING or INNOVATION. If the system that you changed works, you get more points.

If the new system saved money in design, installation, or maintenance, the bigger the number of points you get.

The bigger the changes or the bigger the consequences (more money), you get more points.

After a while, you could trade the points for bonus money, vacation, or other benefits—just like frequent flyer mileage.

Logical Thoughts at 4:00 a.m.

Take a chance— get points.

Just like an airplane— take a chance— ride the plane— get points.

All this can be explained by a basic law of high school physics:

Things don't change unless pushed.

COULDN'T YOU FIX THE DISHWASHER?
NO... THERE WAS NO D/P CELL, PID CONTROLLER, DCS, PLC, OR CONTROL VALVES!
...HEH HEH HEH....
INSTRUMENT JOKES HAR HAR HAR

Spaces

If you start to talk about "spaces" to most people, they look at you like you're a little weird. Space is a place between the earth and the moon, planets, sun, and stars. It is where space ships go. But, in the United States Navy, people know about "spaces."

Every place or room on a ship is a space with a name. The names are not complicated or original. One example is "main space." Spaces on ships are usually not much larger than they really need to be.

If people who lived in houses would follow these two simple rules, some interesting things might develop. Remember the two rules: every space has a name, and the space must not be larger than necessary. Here are some possible consequences:

There would be no "living rooms" with plastic covers on the furniture where kids, dogs, and adults are not allowed to sit down.

The name "basement" would not be allowed. It is not descriptive enough. Some people put the kids in the "basement." Others use the "basement" for parties, washing clothes, a spare refrigerator or freezer, spare bedrooms, the furnace, pool tables, etc. Most "basements" are bigger than they need to be.

"Dining room" is also a bad name. Middle-class women store expensive furniture in this room. The furniture is called a "dining room set" and is used to take up space but is very seldom used. If the table, which is part of every "dining room set," is occasionally used for eating, there is a board covered with some kind of plastic on top of the table. This protects the table but keeps anyone from seeing the beautiful wood grain that cost an extra $3000. Also included in the set are big boxes with doors that have glass and/or grating. Inside the boxes are well-lit shelves with little statues, cups and saucers that are never used, silver candle holders that are never used, and various other things called "knickknacks" that have been passed down from the wife's grandmother. Men never get "knickknacks" from their grandmothers. "Knickknacks" can also be purchased at flea markets or at little shops with fancy names. The fancy shops get their "knickknacks" from flea marketers who get their stuff from the estates of grandmothers who didn't have any granddaughters.

"Dining rooms" also have a small table on wheels that some people call a "coffee table." It may even have an old Russian coffee pot sitting on top; however, no one makes coffee on a "coffee table" and it certainly never gets rolled around on its wheels except maybe to clean some imaginary dust that was about to accumulate underneath.

Kitchens can be interesting "spaces." It is a place where food is stored, prepared, and eaten. There are clocks, TVs, radios, and a variety of methods for cooking, heating, reheating, defrosting, and burning food. A well-equipped kitchen space can contain: toasters, toaster ovens, popcorn makers, ranges, conventional ovens, microwave ovens (with or without browning capability), broilers, radar ranges, and convection ovens. Yet, with all these high tech methods available, why do people go outside in nasty hot or cold weather to cook on dirty grills using dirty charcoal and lighter fluid? Or, even worse, they use a gas barbecue, which may be the epitome of middle-class suburbia. The reason that they cook outside, without the use of a microprocessor, is to keep the guests, dogs, and kids out of the "living room" and "dining room." The party or special eating event is held on the lawn, patio, yard, porch, cellar, basement, or just outside. (These are not the kind of names that the United States

Navy would use. Remember that "spaces" must have descriptive names and must be barely big enough to accomplish their mission.)

Back to kitchens, which are supposed to be used for food-related activities. Some people use the area for talking on the telephone, doing homework, serious and non-serious discussions, bill paying, reading, writing letters and books, fighting and arguing, and worst of all, sticking notes and photos on walls and appliances. When we were kids, no one put notes and pictures on a refrigerator. We think that during the "REALLY BIG WAR" the Japanese discovered that refrigerators were magnetic and that little colored magnets could be made cheaply. They used it as a bargaining tool during the peace negotiations with the condition that the United States not commercialize the idea before they had the opportunity to perfect the "compact" car. When we presented this theory to some patriotic older people, they told us that they didn't need notes to remind them what to do. In addition, they all had photo albums that they received from the hotel where they spent their honeymoon.

More on kitchens: Cabinets under the sinks should be outlawed by the EPA and OSHA, especially if children, pets, and people without Ph.D.s in chemistry are allowed in the kitchen. These spaces may contain any or all of the following:

1. Trash and/or garbage containers with plastic bags
2. Garbage disposal motor
3. Pan to catch drippings from garbage motor
4. Water pipes, drain pipes, water hoses to dishwasher and spray nozzle
5. Pan to catch drippings from #4
6. 7 kinds of liquid cleaning/polishing solutions with a variety of toxic ingredients

7. 3 kinds of solid cleansing/polishing solutions with a variety of toxic ingredients
8. 5 kinds of spray or aerosol cans for cleaning/polishing with a variety of toxic ingredients
9. Scrubby pads, steel wool, rags, and brushes for spreading the ingredients in items 6 through 8
10. One inoperative fire extinguisher
11. Miscellaneous tools consisting of at least a rusty screwdriver with a broken tip and a piece of bent metal to turn a jammed garbage disposal
12. Various insect and rodent deterrents

The potential for chemical warfare from the contents of these cabinets is beyond the capability of Moammar Gadhafi's Libyan poison gas unit.

The United States Navy would never store toxic chemicals in a box used to cover pipes. They never would even bother to cover the pipes.

Kitchens may also be used to store empty bags from the market, bags and containers for storage of anything, telephone books, cook books (even though never used), and the most important pile of unpaid bills.

I GO TO SO MANY MEETINGS, I DON'T HAVE TIME TO DO THE DETAILED WORK WE WORRY ABOUT IN THESE MEETINGS...
WE BETTER HAVE A MEETING TO DISCUSS THIS...
HEY! MEETING!
BIG MEETING
Meeting
BOB

Instrument Engineers Adventurers' Club

There is a small band of fellows (two to be exact) who are bored with the role of being typical instrument engineers. They like to live by their wits and see how much adventure there is to be had in Corporate America. While having fun, they are getting lonely (besides, they need to spread the blame and confuse the opposition). So, they have issued a call for new recruits. In order to qualify, you must work your way up from the Novice to the Daredevil level in one of the following categories. Be careful not to go overboard and become a crazy person. Send proof of your achievements to the author and he will send you a membership card and a secret decoder ring for your boss' memos.

For Performance Reviews

Novice: Have a friend, posing as the president, interrupt the review with a phone call to ask your advice.

Advanced: Hire a prestigious public relations firm to act as your intermediary.

Daredevil: Tell your boss someone is sitting in his or her Beamer, switch badges when he or she turns to look out the window, and swap roles.

Crazy Person: Bring proof of your accomplishments.

For Excitement

Novice: Hide your boss' *Wall Street Journal*.

Advanced: Substitute a copy of the *Journal* from the Great Crash.

Daredevil: Type a notice that all Beamers are being recalled.

Crazy Person: Tell your boss you heard that Beamers have leftover Yugo transmissions.

For Advancement

Novice: Attend all management courses and meetings, give them rave reviews, learn the buzz words, and become a meeting specialist.

Advanced: Get an MBA from Harvard.

Daredevil: Move in next door to your president and invite him/her over to discuss stock options.

Crazy Person: Rely on your accomplishments.

PSST... HEY STANLEY, IT'S ME, GEORGE... YOU BETTER WATCH OUT, THERE'S A FLIGHT OF PECANS COMING DOWN SECTOR FIVE!
MOTHER OF ALL PIES
OH NO! EDWIN GOT HIT WITH A RHUBARB!
PIE SHELTER
I'M HIT! I'M HIT!
OH MY GOD! IT'S ANOTHER SUICIDE SQUAD! LOOKIT! THAT GUYS' GOT PIES ALL OVER HIM! IIIIEEEEEE!!!
HERE COMES THE ARMOR!
BANZAAIII!!!
...BUT I'M A MANAGER! AND I HAVE KIDS... AND AN MBA! DON'T PIE ME!
WHO CARES!? YOU'RE GETT'IN YERS, LOUIE!
HEH HEH... I KNOW THAT'S A BLUEBERRY... MY FAVORITE!
FIRE ONE!!!
ZING!

A Pie in Every Cubicle

Maybe it was the low moral caused by the growing disparity between salary increases of the average employee and the bonuses of upper management, or maybe it was a sense of humor that caused the chief executive officer to proclaim "Let them eat pie!" Sensing he was on a roll, he followed it up with the promise that he would not rest until there was a pie in every cubicle. So for the Thanksgiving holiday, a plan was developed to give away pies.

In the spirit of modern business, the whole effort was contracted out to the cafeteria caterer, who saw it as an opportunity to kill two birds with one stone or, in this case, two engineers with one pie. A free pie for every cafeteria receipt would help curtail waning enthusiasm for their grease bowl delights.

It was an interesting sight to see engineers balancing pies on the way back to their offices. Since there was no minimum dollar amount for the receipt, some astute engineers made several trips to the cafeteria and bought just a cookie or a soda to get another free pie. One particularly gluttonous engineer, known for his ability to locate and raid meeting rooms with free doughnuts, managed to collect a cubicle full of pies.

The engineers might have peacefully carried their pies home that evening if pie day had not been marred by an unfortunate event. The heating and ventilation system failed, and temperatures and, eventually, pies shot up on the west wing, the home of the party association (the sponsor of wild parties and practices fostered during the college experience). The sight and smell of pies starting to rot was just too much stimulation for this group. Eye witness accounts say that one person started the mess by lobbing, one by one, his assortment of pies over his cubicle walls. Rapidly, it escalated from an isolated incident to an all-out war. Engineers donned hard hats and safety glasses and placed scouts at strategic places to warn of incoming pies. Suicide squads were sent to other wings in Rambo fashion with pies strapped on as ammo belts. Several managers sustained direct hits when group members showed up at staff meetings armed with pies. Simulation programs were written to model pie warfare and predict the flavor, freshness, and trajectory so that targets could decide whether to duck or open their mouths. Model error caused rotten rhubarb pies to be mistaken for fresh strawberry pie, much to the dismay of the system section's elder statesmen, who ingested large quantities of the putrid pies before the avalanche of data forced them to admit their program was wrong.

The pie in the sky event ended with an uneasy negotiated peace, with the west wing gaining the right to strangeness and its free expression. Several months later, the west wing was transferred in box cars to a remote outpost in North Dakota.

When the chief executive officer asked how the employees liked the pies, he was told they were excited and physically moved by his generosity, but the offer should never be repeated.

...IN RETRIBUTION FOR YOUR EXCEEDINGLY MEAGER SUPPORT OF LAST YEAR'S DRIVE, I HEREBY SENTENCE YOU TO BE THE SAVINGS BOND PROGRAM SOLICITOR!
ABANDON HOPE!
HAVE A NICE DAY...
INTECH

The Top Ten Terrifying Thoughts of Instrument Engineers Just Before They Fall Asleep

Just before you fall asleep and are caught between the conscious and subconscious world, prospective events can assume ominous proportions. This is an accounting of one such night after a particularly difficult day. The thoughts are arranged in order of increasing scariness.

10. **What if they gave the operators my phone number?** Will I be blamed for every complex control system that doesn't work? Do they expect me to solve problems at 5:00 a.m. that I don't understand at 2:00 p.m.?
9. **Will I have to learn a new keyboard?** What will I do when they change the mainframe terminals, personal computers, operator interfaces, and configuration workstations? Will my mind just go blank, or will I suffer a mental meltdown and babble incoherently?
8. **How do Coriolis mass flowmeters work?** How can I understand something I can't pronounce? If I stand on a record turntable, will

Logical Thoughts at 4:00 a.m.

I get a feel for it before I get knocked off by the needle?

7. **Will I have to learn a new goals program?** Will my automatic goals verbiage-generating program be obsolete? What if I have to write goals for real events that support the idyllic management goals of my boss?

6. **What if the buyer goes out for quotes on my rush job?** Will a request for bids be sought from the third world? Will my control valves be plastic and made in Taiwan?

5. **Is the cafeteria a test site for the company's products?** How do they concoct such weird colors and tastes? Was that the chief chemist I saw coming out of the kitchen?

4. **Could I be attacked by roving bands of project managers?** What if they read the book *How To Become an Instrument Engineer?* What if they become violent? Will they wave knives and force me to say kind things about schedules and budgets?

3. **What if God is a lot like my boss?** Will the ultimate performance review be a lot like last year's? Will my rating depend more on what I didn't do than what I did do?

2. **What if another nuclear engineer becomes president?** Does an actor make a better president than an engineer? Would an actor make a better manager than an engineer? Is my manager an actor?

1. **Will they eliminate the early retirement program?** Will they run out of money just before I become eligible? Will I spend the rest of my life in a cubicle?

Electrons

Where do all the electrons come from?
In the old days, before the revolution, high school chemistry teachers taught us that electrons had protons somewhere near them. This meant that every negative had a positive to balance it out. But, if you go to college and take electrical engineering courses, they never talk about the protons or the positives. All they talk about are all the trillions and trillions of electrons shooting through wires and electronic boxes. You might think that all these electrons are free since there are so many of them. However, anyone who pays an electric bill knows how expensive they really are.

After programming our new computers, we made the following startling discoveries:

1. Electrons cost so much because no one knows what to do with the protons.
2. It's getting harder and harder to find places to store these excess protons; so the price keeps going up. It's like how the price of landfill keeps increasing. If we could find a good thing to do with the trash, we would pay less for the products that cause the trash.
3. It naturally follows that if we could find a good use for protons, the cost of electrons would decrease. With this as a basic premise, we propose to:

    a. Create a field of non-science called "PROTRONICS."

Logical Thoughts at 4:00 a.m.

---

    b. Everything would be opposite to "ELECTRONICS" or 180 degrees out of phase.
    c. The people working in the field would be called "Protrical Developers." They could not be called "engineers" because the mathematics system would be non-existent or opposite.
    d. Calculators, slide rules, computers, and logical thought would serve no purpose and would be forbidden.
    e. Recruitment of workers would come from incompetent mental health professionals, lawyers, and fine arts majors.
    f. The protons would flow backwards from electrons, which means that clocks would have to turn in the opposite direction.
    g. Passing large quantities of protons through conductors (wires) would produce cold instead of heat. Toasters would be freezers. Refrigeration machines would not need compressors to produce cold air.
    h. Proton batteries would work better in cold weather, so starting your car would be easier in the winter.
    i. Proton bulbs would produce light when the switch was turned off. This is what is termed "fail safe." During power failures, all the lights would go on and the furnace would warm the house. If it got too warm, you could open the windows.
    j. The system would be inherently safe because no one could get "ELECTROCUTED."

THE SMOKE SCREEN

There is a theory growing in popularity that the secret ingredient to make electrical circuits function is smoke. This is based on numerous instances where electrical devices, particularly graphic display screens, have stopped working after smoke was sighted leaving the system. It seems the more smoke that escapes, the greater the difficulty in getting the device to work again. Huge belches of smoke probably means the system will never function again.

I TOLD OPERATIONS WE ARE READY TO DOWNLOAD THE NEW CONFIGURATION...
FATAL DISK ERROR
POWER DOWN AND RETURN COMPUTER TO MANUFACTURER
WHISTLE
GEEGAW
BELL
GIMCRACK
Bob

Some Wonderful and Nonwonderful Things Computers Will Do

In the old days, there were tanks of all sizes that needed to be kept full of a specific liquid (mostly water). One way of doing this was to have some sort of measuring device on the tank that would detect when the level was beginning to drop. This device would be connected to a pump and/or valve that would start/open to fill up the tank again. There were hundreds of varieties of this system that varied in complexity and cost. Along came the computer and/or microprocessor.

The possibilities are now limitless. Just recently, we saw a system that worked something like this...

1. The level began to drop.
2. The computer received the signal that the level was dropping.
3. The computer determined how fast the level was dropping.
4. The computer sent a signal to a flow control system to begin adding liquid that had some relationship to the calculation done in step 3.
5. Five minutes later, the computer checked the level in the tank.
6. Depending on how fast the level in the tank was going up or down, the computer did another

Logical Thoughts at 4:00 a.m.

        calculation and sent a new signal to the flow control system.
7. Ten minutes later, if the level was still dropping, the computer would begin reducing the flow to the users.
8. And on, and on, and on—limited only by your imagination.

How about a computer that designs a bridge that doesn't FREEZE BEFORE THE ROADWAY?

Or a computer that translates ideas/words of one language into ideas/words of the same language so that everyone understands the message?

HEY DUDE... WHERE'S THE HOSPITALITY SUITE?
ZZZZZZZZ
CONTROL VALVE
TYPE: GREEN

How About Giving a Talk?

One day the phone rings and an unfamiliar voice begins to tell you a story about how his organization is having a technical symposium next year in some remote location. Since you are an expert in something or other, he would like you to give a talk. That's what you hear first. If you are an experienced speaker, you will ask for more information in writing. If the person on the other end of the phone is convincing enough, you may commit to the talk with some very sketchy information. Here is what the real story is...

Mister Nostalgic N. Smith is a retired engineer from a large company that offered him a golden handshake. His wife doesn't want him around the house all day and he misses feeling important, so he goes to work for a so-called "not for profit organization" that is promoting the salvation of the world through the increased use of Sight Flow Indicators with Spinners (commonly known as SFIS). Mr. Nostalgic (calling him Smith may confuse him with someone else) doesn't really know much about sight flow indicators, but he's kind of a nice guy that people hate to say "no" to.

Mr. Nostalgic is offering you the opportunity to address this distinguished organization at their annual technical symposium in North Dakota at the beginning of February. When you ask why anyone would have a meeting in North Dakota at the beginning of February, he responds that the winter is over in January and it's an opportunity to see a wonderful part of the United States.

Logical Thoughts at 4:00 a.m.

He carefully explains that you were selected to give this important presentation as the single representative of the Petroleum, Chemical, Pharmaceutical, Aeronautical, Beer, Electrical Power, and Gas Transmission industries. (Sounds awesome—except that he couldn't get anybody else to come.)

His organization would like you to explain what all the industries that you represent want from the Sight Flow people during the next century, so that they can plan their research and production budgets. In order to make the offer more attractive, the organization will pay for your hotel and lunch. In addition, one of the executives may buy your dinner, but he really can't commit to that.

Since Mr. Nostalgic is such a nice guy and giving talks enhances your reputation within your own company, you agree to the presentation. You write the paper, get approval to deliver it, and wait for the event to occur. About three weeks before the meeting, you accidently see an advertisement inviting any interested purchasing agents, company executives, or potential users to this very informative gathering. There is also a description of classes that will be taught to the uninformed by experts in the field. One of these classes describes how sight flow indicators are applied and misapplied, and you are the speaker.

The class is scheduled for 4 p.m. on Friday afternoon, which makes it the last talk of the session. This means that most people will have gone home and you will miss the last bus to the nearest airport. When you try to reach Mr. Nostalgic (who works out of his house during the winter), you find out that he spends December and January in Florida.

Within a few days, a letter arrives congratulating you for speaking and informing you that a luncheon for the speakers will be held on Monday; however, since your talk is on Friday, only Thursday night's hotel room will be

How About Giving a Talk?

paid for you. Your spouse can stay in the room with you at no additional charge. (You know that the Sight Flow Indicator people are progressive by their use of the word "spouse.") Their organization brought in a management consulting company that taught a class on the Diverse Work Force, which taught them how the minorities will soon be the majorities, and why people on the east and west coasts do weird stuff after work. Your inclination is to back out of the commitment as tactfully as possible, but, in reality, withdrawing at this late date will make you look like a "jerk." So, off you go.

A badge with a fancy ribbon awaits you at the reception desk. A person with a more elaborate ribbon than yours asks you if you brought 200 copies of your talk. You inform this person that the paper was sent two months before and that it was your understanding that they would reproduce it. After a check of the files, they show you that your paper was not in a proper format and therefore could not be reproduced for the proceedings.

At all sessions before lunch, you notice that all the speakers are using slides. In your briefcase is a beautiful, professionally made set of colored overhead transparencies. When you ask about an overhead projector, they carefully explain that the letter that showed you the required format for paper also explained that, since the meeting room would be large, only slides would be acceptable. However, since you are such a dum-dum who doesn't read his mail, they will go to the great expense and inconvenience to see if the hotel will rent them one. Fortunately, one is found in a back closet, but both the bulbs are burned out. A search finds a bulb with the proper socket but only half the wattage. When lunch and the breaks run late, your 4 o'clock session starts at 4:35. By 4:45, three quarters of the chairs are empty and people are leaving faster than you can get through your transparencies.

Logical Thoughts at 4:00 a.m.

At home, your spouse is angry that you went away for three days on this dumb trip. At work, your boss asks why you stayed an extra night in such an expensive hotel. Weeks later, you still ask yourself why you ever agreed to give the talk, why you didn't back out when things started to go wrong, and why you didn't tell those idiots what was wrong with their entire operation. The remaining question is, "What will you tell a Mr. Smith the next time he calls about presenting a paper?"

...O.K., THIS IS THE PLAN... WE TALK LIKE M.B.A.s, POSE AS A PRESTIGIOUS CONSULTING FIRM, AND RECOMMEND THAT THE BOARD OF DIRECTORS HAVE A MAJORITY OF ENGINEERS BECAUSE ENGINEERS WILL BE THE KEY TO BIGGER PROFITS...
BIG KAHOONA
THE BIG CHEESE

Believe It or Don't

Sometimes statements are made that just seem too radical to be true. Here is a collection of expressions from unknown sources. It is up to you to decide whether they are even remotely possible or are just fig newtons of the author's imagination.

1. A business student with a Master of Science in Engineering got a higher starting salary than one with a Master in Business Administration.
2. A business student switched to engineering because the business courses were tougher.
3. An engineer was elected president of a chemical company. Wait, that's too bizarre for even this book! How about an engineer was elected to the board of directors? You don't buy that either? Would you believe an engineer was selected to be a vice-president? You say that's not conceivable? Well, how about this? An engineer was promoted to a position that impacted the company's future. No? Would you consider an engineer was once asked for advice from the top management? What, that's not possible? Well, OK, but several authorities claim a president once asked an engineer for directions to the rest room.
4. An MBA understood and appreciated the contribution of an engineer.
5. An engineer understood and appreciated the contribution of an MBA.

6. A company said, "We must increase our manufacturing base instead of selling assets to increase the return on capital."
7. A company declared, "The heck with what Wall Street says, what we really need are plants engineered to outperform the competition."
8. Tokyo companies express concern about the low ratio of MBAs and lawyers to engineers in Japan.
9. A company affirmed that employee career growth and self-fulfillment were more important than stock price growth and extra dividends.
10. Russia's top officials pleaded with the U.S.A.'s President to send over surplus lawyers to help their struggling economy.

LOST KITTY
REWARD
CAR FOR SALE
Pyromaniac Special
...with FREE FIRE
EXTINGUISHER and
FIREMAN's HAT!

How Many Fellows Does It Take to Pick Up a Recalled Car?

When this Fellow was notified that his sportscar was recalled due to fires, he was surprised because he had not seen so much as a puff of smoke in four years of service. But, he reasoned, the manufacturer must know what's best, so he left the car at the local dealer for the recommended fix while he went on a business trip.

Upon his return, the car was ready but covered with six inches of snow. He was dismayed to find out his car wouldn't start and astonished when he noticed smoke coming from the rear lid vents. In this wonder work of Detroit, the rear lid covered the engine compartment and a luggage compartment designed for elves.

The Fellow decided that the moment required quick action and that a study of the recall order and a test by dynamic simulation was not in his best interest, especially since the car was made of plastic. He ran to the back of the car, slipped and fell flat on his face, pulled himself up on the rear bumper, unlocked the pseudo trunk, supported the snow-laden lid with one hand, and threw snowball after snowball at the fire with his free hand (it is amazing how fast snow melts in a fire).

When he momentarily let loose of the lid, it slammed shut from the weight of the snow. The full implications of the crashing lid were not known until the fellow returned with a rather excited service manager with a fire extinguisher who asked the Fellow to unlock the trunk.

Logical Thoughts at 4:00 a.m.

In his haste, the Fellow had dropped the keys in the luggage compartment. The look given by the service manager was something less than respectful of the Fellow's credentials.

Fortunately, the parts department could cut another key. Unfortunately, the parts department doesn't like to cut keys, so the guys behind the counter pretend the Fellow didn't exist for at least an hour to make sure he realized the full consequences of his actions. So the Fellow waited and reflected whether the fire was really out or whether his sportscar was destined to become Silly Putty.

The fire was out, but the dealer refused to furnish a courtesy car, due to some uncertainty of obligation, until the Fellow reminded them the car caught fire on their lot after being serviced on a recall to prevent a fire.

Two years later, another recall was issued to prevent fires. The Fellow didn't know whether to take his car in but decided if he did, he would bring redundant flame scanners and a fire extinguisher when he picked up the car and that it would require ten fellows—one to hold the lid, one to hold his keys, two to set up the flame scanners, one to operate the fire extinguisher, and five to read the instruction manuals.

The Uffda List

Uffda is a Scandinavian term that defies exact translation. It is a heart-rendering expression of one's feelings at particularly poignant moments. Everybody has had such occasions in their lives. Some days you may feel as though you have had more than your fair share. The following list is dedicated to all of you who have had one of those "uffda" days.

1. Uffda is a rainstorm in rush hour traffic. A double uffda is a snowstorm at 5:00 p.m. on Friday.
2. Uffda is a new computer program to do a task that takes five minutes with a pencil. A double uffda is a full day of instruction to explain the new system.
3. Uffda is trying to find new words to say the same old things in your goals document for twenty years. A double uffda is a new boss who takes it seriously.
4. Uffda is a computer program with its own set of undocumented special function keys. A double uffda is a key template with unrecognizable nomenclature and no exit key.
5. Uffda is a deadline for detailed work and a week full of meetings. A double uffda is a meeting for you to explain why the work is not done.
6. Uffda is a four percent raise when the inflation rate is five percent and the company has had its best year. A double uffda is the president's bonus of a million.

7. Uffda is a computer system that goes haywire and a support person last seen two years ago. A double uffda is a self-service manual pinned to a wall and about to fall due to excessive weight.
8. Uffda is a cubicle next to a person who talks to himself. A double uffda is a cubicle next to a person who answers himself.
9. Uffda is a cubicle next to a person with a hearing problem. A double uffda is a cubicle next to a project manager with a hearing problem and an impossible schedule.
10. Uffda is a pipe fitter doing instrument calibration. A double uffda is an operator with channel-locks giving an instrument a whole new look.

...IF I COULD JUST GET THAT SMOKE IN SOME WIRES, I COULD INVENT THE TELEPHONE!
RING MY BELL
?
alex

Phone Tag for Fun and Profit

It is extremely rare for two human beings to communicate by phone. The proliferation of the automated answering systems and the dawn of the meeting specialist mean that the probability of actually contacting a human by phone has approached zero. Now that computers place calls, we can end up with computers talking to computers, computers making major decisions without human intervention, and the "Hal" of the business world taking over Corporate America. (Have you noticed how the computer rooms are nicer than the cubicles occupied by the people who supposedly run the systems?) Humans still think they are in control and don't even realize they have lost the ability to communicate in real time (engineers, programmers, and scientists don't mind because they basically don't like to talk to people—they are too unpredictable). What is worse, most insist on functioning in the same old way where no change proceeds until it is confirmed by verbal communication. The result is phone tag and stagnation. Here are some of the rules on how the game is played.

1. Screen all your calls by use of the answering system. This will ensure no one can ever talk to you live and ask embarrassing questions like "Why doesn't your recommendation work?" until you can thoroughly investigate the problem and come up with a good excuse.

2. When you have tallied a long list of messages, wait until you are about to leave for lunch or another meeting and phone all your replies. This guarantees your line will be busy or you will be gone if someone actually tries to expeditiously return your call.
3. Never update your greeting to inform people of extended meetings, business trips, alternative contacts, or return dates. Above all, don't tell people of your early retirement. This will increase the number of messages (especially blank ones from frustrated callers who needed you to make a decision). The standard "I am away from my desk" will help increase the tedium of the game and the insanity of the players.
4. Never leave a detailed message when you get someone else's answering system. This means you can play phone tag for months without even knowing why. Some of the more interesting rounds of phone tag were started by a wrong number.
5. If you ever actually talk to a human, say "You are sure hard to get hold of." This will help nurture a feeling of innocence.

Imagine what would happen if somebody with a distorted sense of humor (certainly not the author) would call everyone in the directory and leave the phone number of someone else in the directory. The resulting rounds of phone tag might become self-sustaining and all productive work might cease. Hmm—maybe it has already happened.

Gerry's Idioms

One of the sad things about a mature (another word for old) engineering department is the departure through early retirement of its more remarkable employees. Most of these graduates have said memorable things that were never documented and will be forever lost. What follows is a meager attempt to save for posterity the paraphrased expressions of one prospective retiree.

"All concrete breaks up, it is just a matter of time. Why not use gravel and get the end result early?"

"If every Fellow got the office window he or she wants, it would leave the rest of engineering in the dark."

"In twenty years, none of this will matter."

"The only reason managers and experts haven't hit into a triple play is that there are already outs or not enough people on base."

"An instrument engineer reaches full maturity after realizing there is no correlation between an automatic butterfly valve spec and installed performance."

"The only thing worse than command line software is a menu-type interface designed by a programmer who wants to use the command line."

"Redistribution of wealth has only been dreamed about by politicians. However, real accomplishments have been achieved by manufacturers of mass flowmeters."

"It is essential to commit to a Distributed Control System (DCS) so that the vendor can pass on his or her expectations of you, tell you what else you need to make it work, fully explain its limitations and deficiencies, and assist in redefining the process to fit the control system."

"DCS reference manuals can't be identified by letters because of our skimpy alphabet."

"Some humorous aspects of post-project critiques have been the original DCS schedule; money saved by Computer-Aided Design; three-dimensional modeling; contractor implementation of verbal instructions; unique contractor interpretation of written instructions; controlling an undocumented process based on undefined research; teaching English to a digital person; and explaining control terminology to a DCS programmer."

"A fault tolerant (triple redundant) system shall cost between a simplex system cost tripled and a simple system cubed. The system shall be housed in one file so as not to require three faults, fires, or surges to disable it."

Bad Rap

With the extreme popularity of rap among teenagers, there is an opportunity to reach out to the prospective instrument engineers of tomorrow through a rap song about one man's career in instrumentation. The potential effect is enormous when you consider how often and how loud rap music is played (certainly in violation of OSHA noise limits and audible to anyone within a ten-mile radius) and how certain rap groups have become household words. The following are the lyrics of the lead track from the soon to be released "Two Live Fellows" album. Even though the words are clean, you might get your son or daughter's attention by pounding a rap beat on the table and shouting the lyrics. Here is your chance to bridge the generation gap and show your kids that you are alive and that engineers are people, too.

> I started the show
> twenty-two years ago
>
> I emerged from my college years
> with no real ideas
> of what the heck I could be
> with an engineering physics degree.
>
> But I was on the capitalism scene
> ready for the corporate machine.
>
> I asked the recruiter, "What do they do, Bro?"
> He replied, "I don't know.

Logical Thoughts at 4:00 a.m.

"But the job pays good bread
and you've a way to get ahead.

"This is no free meal
but it's no bogus deal."

I accepted the offer with excitement
and tried to fit in with the establishment.

I was surprised to find you never designed something yourself
but always picked something from unknown catalogs off the shelf.

When I was given a chance to specify a lot
and left free to give it my best shot.

I was literally and figuratively moved
until I learned the project would never be approved.

I needed to know a thousand details or more
on instruments I had never seen before.

My manager knew my college education
did little to prepare for this situation.

So he gathered up leftover instruments and peers
and set up a program to train new engineers.

But my dog ate one of my contacts
like it was one of his tasty snacks.

So I started a company school
looking like a one-eyed fool.

Maybe it was my creative participation in the lab destruction that convinced my boss to ship me off to field construction.

I wandered the construction site day and night
and made sure the instruments were installed just right.

I learned how to accomplish a start-up or shutdown
with the latter usually accidentally found.

These years gave me the confidence
to solve problems of significance

And a real-world mentality
to ensure a system's practicality.

So when I moved to Process Control Technology,
I used a down-to-earth methodology.

The analytical criticism of my colleagues was so intense
I had to be technically astute in a larger sense.

I had the chance to spout creative ideas
without political repercussion fears.

The freedom of expression and matriculation
was essential to my self-realization.

I discovered all kinds of technical stuff
but saw technical writing as not interesting enough.

For what good are documented principles to heed
if they're told in a way too boring to read?

So I tried to add some humor
in a way never seen before.

Logical Thoughts at 4:00 a.m.

The result were articles about Wally and the Beave
detailing stuff that would normally make you heave.

The book *How To Become an Instrument Engineer*
was destined to appear.

And the sequel *Logical Thoughts*
became a possibility of sorts.

...PUT THAT BACK, YOU CAN READ YOUR GRANDMOTHER'S COPY!
BUT SHE'S ALREADY COLORED ALL THE PICTURES!
SALE!
INTECH
WOMAN'S Day
Reader's Digest
LOGICAL THOUGHTS AT 4:00 AM

Top Ten Reasons for Buying This Book

On the average, each copy sold of the last book, *How to Become an Instrument Engineer*, was borrowed and scanned by five other people. While complimentary, this behavior is inconsistent with capitalism and the authors' goal to appear on the program, "Lifestyles of the Rich and Famous." But the authors did not want to be obtrusive, so this advertisement is buried deep in the book's contents to convince readers to buy copies. Remember, they make great gifts for the whole family.

10. It exercises both hemispheres of the brain, and it is well known that daily exercise is the key to a long life.
9. You can make your company's restroom an entertainment center with strategically placed copies.
8. If sales fall off, our books will never make it to the racks at the grocery checkout lanes.
7. If you don't, the authors will have to go back to telling instrument jokes to their families.
6. To ensure that ISA executives can go to neat places.
5. So the authors can write off swell lap top computers.
4. Laughter is the best medicine (this is not meant to imply our readers are sick people).
3. Our boss' door is open. He is asking us what's wrong. He even asked us to dinner. We are on a roll— don't stop us now!

Logical Thoughts at 4:00 a.m.

2. It was panned in the *Harvard Business Review*.
1. The Japanese would never write such a book.

...WHAT'S NEXT?
HMM... MAYBE A SCRIPT FOR A T.V. SERIES ABOUT ETHICS IN A CONTRACT ENGI-NEERING FIRM CALLED L.A. ENGINEER...